Song of the Scythe

Bruce Mawdesley

Song of the Scythe

wood-engravings by Miriam Macgregor

The Whittington Press

Printed and published by
The Whittington Press, Manor Farm,
Andoversford, Gloucestershire

First published by The Plum Tree Press
in an edition of 40 copies

0 904845 57 5

ONE of the loveliest sounds of the countryside today is the music of stone on steel as a scythe is whetted, but alas it's becoming increasingly rare. It was commonplace not so long ago, and from June onwards somebody would be using one in field or churchyard, or at the roadside, but this has now become the exception rather than the rule. If most of the sweat has been removed from man's toil in summer's heat and haze, then so too have the peace and serenity that were part of the old way of life.

The tools of our ancestors were possessed of strength and simplicity, yet retained a natural beauty of line and purpose for the tasks for which they were designed. They might appear clumsy to the untrained eye, yet they were almost as supple as the limbs that weilded them. They became, in fact, extensions of those same limbs, and none more so than the scythe.

Prehistoric man slowly became aware of the possibilities of farming and began to harvest the wild grain and grass which he found growing in abundance. To reap these crops he adapted his hunting tools to husbandry by binding a sharp flint into the cleft of a stick. Thousands of years later he was using bronze and iron, and by a process of trial and error arrived at the conclusion that a curved blade both cut and gathered the crop more efficiently, and by adding a longer handle he discovered that his efforts were less tiring.

The first evidence we have of the scythe is from the culture of the Iron Age, although we cannot be certain whether this was an innovation or merely an extension of the long-handled sickle. The scythe, however, had come to stay, and up till the Industrial Revolution was to remain master of the harvest field.

The components of the scythe are the steel blade, the curved wooden pole, the general term for which is 'sneath', 'snead' or 'snath', these varying according to the district; and two short wooden handles called 'nibs' or 'doles'. These handles are fixed to the snath by iron rings and are adjustable. The snath is usually wider at the base than at the top and was usually picked individually from young ash saplings, but the modern factory-made article is of hickory.

The blade can be adjusted outwards or inwards to cut a wider or narrower swath and is locked into position. The 'grass nail' is a thin metal rod or wire which connects the blade to the snath at the front and serves the dual purpose of taking the strain when the scythe is in work, and preventing the grass from clogging in the angle behind it.

The choice of a scythe is somewhat limited today for there is really only a factory-made tool available, but it is still possible to acquire an old one in good condition, which is far better, for the modern snath tends, in my opinion, to be unnecessarily heavy and the blade doesn't keep its edge for long.

Sometimes on the reverse side of an old blade one may find the initials, Mon, Tues, Wed, Thur, Fri, which stood

for the day of the week on which the metal had been forged, and an old scythesman told me that he would always pick a Thursday scythe, the reason being that by mid week the furnaces were at full heat, whereas on Monday they were sluggish, having been freshly lit, and on Friday they were allowed to start cooling down again.

A good 'rub' or 'whet stone' was kept attached to the belt for the blade must be kept as sharp as a razor to do any good. Sandstone was much favoured before the advent of carborundum.

Up till the beginning of this century most of the scythesmen in the busy haymaking season were Irish. They came across the sea in their hundreds and could be found strolling down the lanes of England with their scythes across their backs, on their way to the farm of their choice where they would return year after year to offer their services. They were a fine body of men and expert scythesmen, and on the evening of their arrival they could be seen setting up their scythes and making practice sweeps on the grass outside their lodgings, which was generally the barn.

Once the work began in earnest there might be as many as three or four men to a field, with the leader, or 'Lord' as he was called, setting the pace with the first swath and the others following. They would start at dawn, when the dew lay heavy on the grass, making it easier to cut. They would generally lie up at mid-day, when the sun was at its zenith. It was customary for the farmer to provide plenty of ale and this in no way impeded the work for a great deal of sweat was lost and had to be replaced because of the very real danger of dehydration. A good man could cut an acre a day.

Young lads were always taught by an old master and the initiation took place on a wide open, level piece of ground. It might well have been called a dancing lesson for rhythm and discipline are essential. The left foot moves slightly forward as the blade is swept in a curve through the crop, its weight resting on the ground. At the end of the first sweep the right foot follows up as the blade is swung back ready to repeat the whole action again. The body pivots slightly with each movement. Like learning to ride a bicycle, once confidence is gained and balance mastered, the scythe is suddenly no longer awkward but seemingly comes alive and moves of its own accord.

When sharpening, as one must do frequently, the scythe is turned upside down so that the narrow end of the snath can be lodged against the instep of the right foot, whilst the blade is grasped firmly by the left hand in the centre of the non-cutting edge. The right hand is then free to use the whet-stone but the elbow must be kept pinned firmly to the side. Never use a broken stone as I once learnt to my cost.

The scythesman does not have to be particularly strong, contrary to popular belief, for their is no real strain or hardship involved once the art is mastered.

Until the mechanical mower was invented lawns as we know them today were little more than rough green plots, kept reasonably short by the scythe or, as with the large parklands, grazed by sheep.

Children would always show great respect for the wicked looking blade as it hung high up on its peg in barn or shed out of season, and rightly so, for it is the simplest thing in the world to remove a foot or a finger by 'messing about' with a scythe. During many a civil war, when the cry had gone round for the peasant to take up arms, farm tools were the only weapons at hand, and along with the bill-hook and pitchfork many a scythe was carried into the fray, although I can't help feeling that many a friend as well as foe must have been struck down.

Rust on a scythe blade is not a bad thing and many were left out for short periods hanging from the crook of an apple tree, out of the reach of children, until they were wanted again. It is a curious fact that this will put a better edge on the blade.

It is sad that the scythe is better known today as a symbol for Old Father Time; at all events it has served Man faithfully over the centuries and seasons, and so I think that is not a bad epitaph. I shall continue to keep mine in good trim and constant use.

650 copies hand-set in 14-point Caslon type and printed at Whittington Court on Zerkall mould-made paper. 35 copies are bound with a set of proofs of the engravings.

FEBRUARY 1983